Learning about Weather

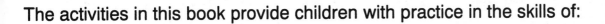

Learning about Weather contains:

- Teacher resource pages

- Student activity sheets

- Picture cards
  Reproduce these cards on tag or cardstock, laminate and cut them
  apart to create a picture file to use with the resource pages in this book.

The activities in this book provide children with practice in the skills of:

- observation
- communication
- making comparisons
- ordering
- categorization

Concepts covered:

- There are many kinds of weather.
- Wind is moving air. It can do work. It can cause harm.
- Rain is water in the air that falls to the Earth in drops.
- Snow is frozen water in the air that falls to Earth as lacy flakes.
- The sun warms the Earth.
- Weather affects how we live:
    clothing
    homes
    activities
- We can see the changes in weather.
- Weather/climate is different in different parts of the world.
    hot and cold
    wet and dry

D1308979

Evan-Moor
HELPING CHILDREN LEARN

Concept
There are many kinds of weather.

Activities:

1. Brainstorm Weather Words
Make a giant cloud on white butcher paper. Write all of the weather-related words your students can think of on the cloud. Add a picture by each word to help non-readers become familiar with them. You may use the picture cards on pages 42 and 43. Leave the chart up throughout your study of weather. Add to the list as your students learn new weather words.

2. Seasons of the Year - A Poem
Learn the season song on page 3. Have children fold a sheet of large drawing paper into four boxes. They then write the name of a season in each box and draw an appropriate illustration for that season. Discuss possible pictures with them before they start the activity.

3. Act Out the Weather
Have children act out movements for different types of weather.
For example:

plopping raindrops
falling snowflakes
wind-blown leaves
sparkling rays of sunshine
swirling tornado
bouncing hailstones

4. Weather Words Activity Sheet
Reproduce the weather words sheet on page 4 for each child. Read the words with the children before they do the lesson. Have them trace the words and color the pictures.

5. A Little Weather Book
Each child will need six 6" (15 cm) squares of construction paper. Have children draw one picture representing a type of weather (snow, rain, fog, etc.) on each page. Then have students write the name of the type of weather on each page. Staple the pages together to make a little weather book for the child to take home and share with his/her family.

Seasons of the Year

Summer, Autumn, Winter, Spring
Different weather they all bring.
Summer showers, Autumn breeze,
Snowy winters, Spring's new leaves.
Summer, Autumn, Winter, Spring,
Different weather they all bring.

Cindy Lunsford

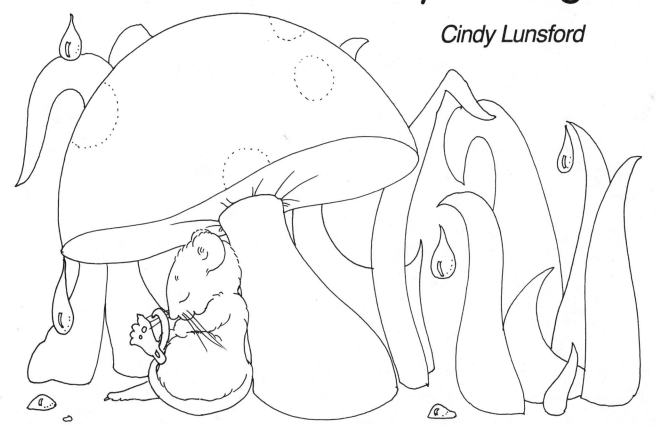

Name

Weather Words

wind

rain

snow

cloud

sunshine

rainbow

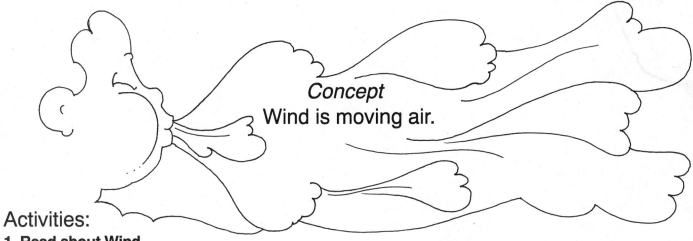

Concept
Wind is moving air.

Activities:

1. Read about Wind

Read *Feel the Wind* by Arthur Dorros (Thomas Y. Crowell, 1989) to your children.
Discuss how the wind feels when it is blowing. See how many different types of wind
they can describe. Have them act out the different types of wind (breeze, hurricane, etc.).

2. Questions about Wind

Ask children to answer questions such as...

"What is wind?"
"How do we know it is all around us when we can't see it?"
"How can you tell which way wind is blowing?"

3. A Wind Scrapbook

Have children collect all the pictures they can find that show wind is blowing. Put these pictures
in a scrapbook. List words or phrases under each picture that describe what the wind is
doing. Put the scrapbook in your classroom library for everyone to share.

4. Wind at Work

Explain that because it is moving, wind can do work. Ask children if they can think of a way
wind does this. Guide them with questions if they are having difficulty coming up with ideas.

"How do sailboats move?"
"How can the seeds of a plant move from one place to another?"
"Can you think of a toy that needs the wind to move?"

Reproduce page 7 and have children work in
pairs to decide where the wind is "working"
and where it is not.

5. Wind Toys

Ask children to think about toys that need the wind blowing in order to move.
Make a list of these toys (kites, sailboats, pinwheels, whistles, etc.). Have them
explain how the wind makes each toy move.

a. Bring in a supply of toys including some that need moving air to work.
Have children decide which toys need wind to work and which do not.

b. Sailboats (instructions on page 8)
Bring (or have children bring) sailboats
to class. Fill a large tub or wading pool
with water. Have children put the boats
in the water and figure out how to provide
"wind" to make them move (use a fan, blow
into the sail, wave your hands at them).
Try out the methods they come up with to
see which work.

If you do not have access to toy sailboats, follow the directions on page 8 to make some.
Guide children through the steps.

c. Pinwheels (instructions on page 9)

Bring in colorful pinwheels for children to play
with. Have children try to figure out two ways
to get the pinwheels to work. Or make pinwheels
using the pattern on page 9.

Give each child a copy of the form on construction
paper. Have them decorate their pinwheels with
patterns or symbols to represent an element of
weather (cloud, raindrops, sun, rainbows, etc.).
Provide a paper punch, straws (preferably plastic)
and brass paper fasteners with long "legs."

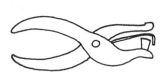

6. Kite Weather

Teach the little wind poem on page 10. Paint
kite pictures to illustrate wind at work.

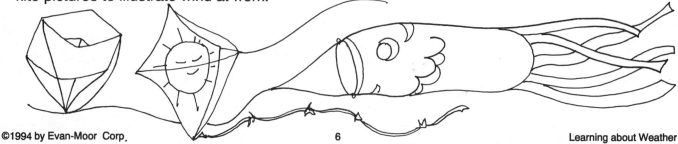

Note: Have children color pictures where wind is working. Have them put an x on pictures where there is no wind.

Name _____

Wind Can Work

Make a Sailboat

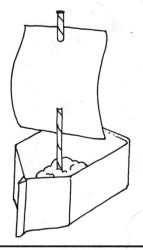

Put out a supply of materials such as these:

- For boat - small milk cartons (side cut off)
- For sail - 4" square pieces of construction paper
- For mast - drinking straws or ice cream sticks
- clay
- glue
- scissors
- hole punch

1. Have children pick up one item for each part of their boat.

2. Give them each a sail pattern (see below) and a lump of clay.

3. Have children assemble their boats as follows:

   a. Put the lump of clay in the center of the bottom of the boat.
   b. Glue the sail to the end of the mast or punch two holes and slip straw through.
   c. Stick the other end of the mast into the clay.

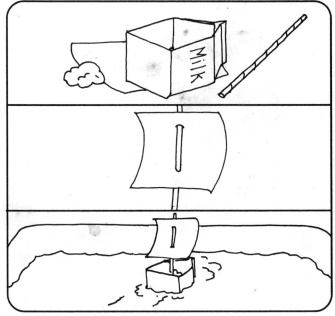

4. Allow time for children to sail their boats in a tub of water. Remind them to pretend to be the wind and blow into the sail to make the boat move.

5. After everyone has had a chance to sail their boat, you may want to discuss which boats sailed best.

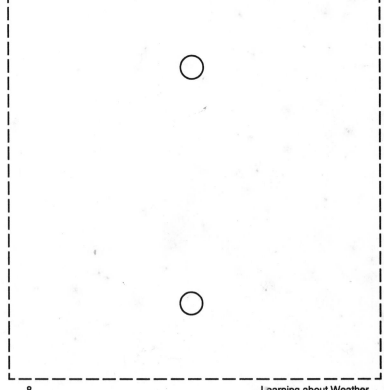

Make a Pinwheel

| Color | Cut out | Punch | Put together | Blow |

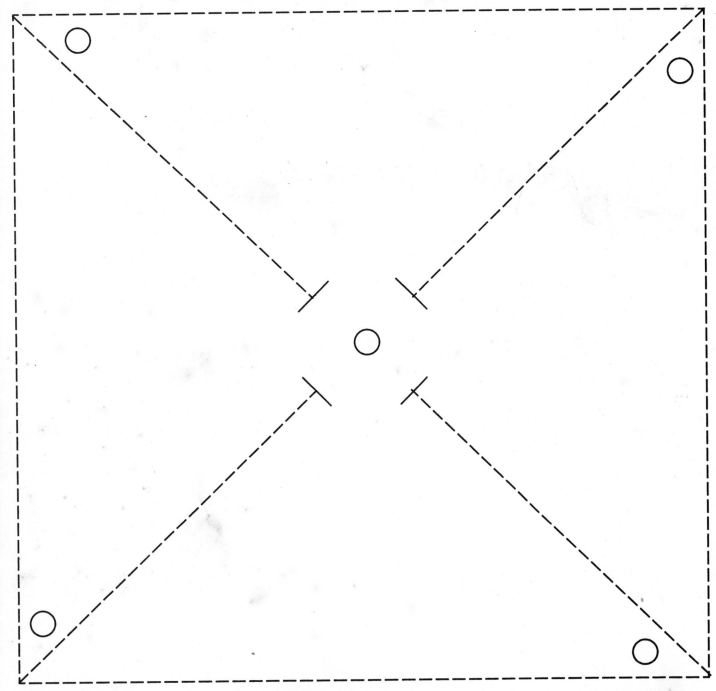

Note: Bring a real kite to fly. Spend some time on the playground on a windy day sharing the experience with your students.

Kite Weather

It's blowing, it's blowing.
The wind is moving by.

Grab your coat; run outside
And fly your kite up high.

Jo Ellen Moore

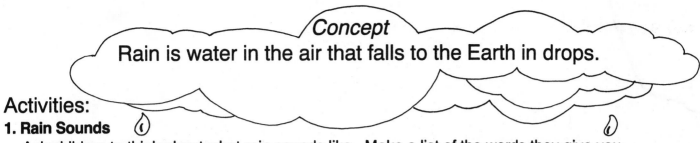

Concept
Rain is water in the air that falls to the Earth in drops.

Activities:

1. Rain Sounds
Ask children to think about what rain sounds like. Make a list of the words they give you.

Read *Listen to the Rain* by Bill Martin, Jr., and John Archambault (Henry Holt and Company, 1988). After children have heard the story, have them add to the list of rain sounds.

Let each child select one of the words or phrases and illustrate it. Put the illustrations into their own class book "Listen to the Rain."

2. Rain Songs
Teach your children the rain songs on pages 16 and 17.

3. Rain Experiments - Condensation and Evaporation
Warm air and water vapor combine to form raindrops. It is difficult for children to understand that there is always water moving in and out of the air. These two activities can help them see what is happening.

Evaporation
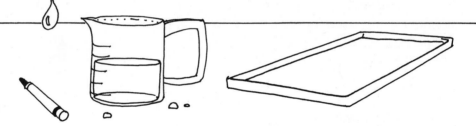
You will need:
- shallow pan
- marking pen
- water
- measuring cup

a. Measure the amount of water you put in the pan. Mark the water level on the side of the pan. Record the amount of water you poured into the pan. Discuss what the children think will happen.

b. Put the pan in a sunny place. Leave the pan for two or three days, observing what is happening to the water. After a few days, measure how much is left in the pan. Discuss where the water has gone.

Condensation
You will need:
- a jar
- lid
- ice cubes
- paper towels

a. Put the ice cubes in a jar. Put the lid on the jar. Dry off the outside of the jar with paper towels. Let the jar sit in a warm place. Observe what happens to the outside of the jar. Discuss where the water came from.

4. Read about Rain
Read *That Sky, That Rain* by Carolyn Otto (Thomas Y. Crowell, 1990) or *What Makes it Rain?* by S. Mayes (Usbourne, 1989) to help children understand how rain occurs.

5. A Rain Wheel

Another way to help children understand how rain is a part of the water cycle is to make "rain wheels." Do the experiments on evaporation and condensation, then have children make their rain wheels. Use the wheels as you review what happens to water as it goes up into the air and falls back to the Earth.

Reproduce the forms on pages 18 and 19. Give each child a brass paper fastener. Put the wheels together as shown below.

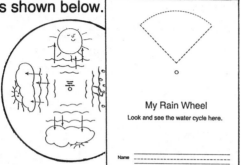

6. Clouds

Explain that clouds are made up of little drops of water high up in the sky. In dark clouds, the drops of rain are closer together and bigger. When the drops get big enough they fall to the Earth as raindrops.

a. Give each child a sheet of white paper. Help children fold the paper into four parts. Show them how to draw a cloud in each box. Then demonstrate how to change each cloud to illustrate how the cloud changes color as it fills up with water drops.

If drawing the steps is too difficult for your students, reproduce the cards on page 44 and have them cut the pictures apart and sequence them in order from white to rainstorm on a sheet of blue construction paper.

b. Read *It Looked Like Spilt Milk* by Charles G. Shaw (Harper and Row, 1947). Discuss how in the story the clouds look like other things.

Make your own version of *It Looked Like Spilt Milk*. Give each child a piece of blue construction paper, glue, a black crayon, a cotton swab and cotton balls. Brainstorm to think of the kinds of cloud pictures they can make.

- Make a light pencil outline sketch.
- Spread glue inside the area using a cotton swab.
- Stretch the cotton balls apart and lay the cotton on the glued area.
- Write a word, phrase or sentence to describe the cloud picture.
- Let the pictures dry thoroughly.
- Staple them into a cover for a class book.

7. Rainbows

a. Use a prism to create a rainbow anytime you want in the classroom. You may want to have several so small groups of children can enjoy exploring with them. Teach them the rainbow chant on page 21 to help them remember the order and names of the colors of the rainbow.

b. Read to your children the rainbow poem on page 20. Then take your class outside on a sunny day. You will need a garden hose with a sprinkler attachment that produces a fine spray. Stand with the sun behind you. Fill the air with a misty spray. Have children move about until they can see the rainbow colors. (Your eyes need to look at about a 45-degree angle.)

c. Paint large rainbows on sheets of butcher paper. Be sure to have out all seven colors of the rainbow.

d. Use an overhead projector to show how new colors are created by mixing red, blue and yellow in different combinations.

You will need:
- clear plastic cups
- red, yellow and blue food colors
- plastic spoons (for stirring colors)
- overhead projector

Mix water and food coloring to create a cup each of blue, red and yellow water. Set these cups on the overhead projector. Have your students name each color.

Explain that something is going to happen when you mix two of the colors. Place an empty cup on the projector. Pour some of the yellow water into the cup. Add blue water a little at a time until you have formed green. Follow the same steps with yellow and red to make orange and with blue and red to make purple.

         Learning about Weather

8. Slicker Kids

Reproduce the form on page 15 for children to use to make children dressed in rain gear.
(If you feel your first graders are ready, have them cut out their own slickers, boots and hats from construction paper instead of using the form.)

Each child will need these materials:
- form from page 15
- 9" X 12" (22.8 X 30.5 cm) piece of white construction paper
- crayons
- dry blue tempera paint in a salt shaker type container
- a spray bottle full of water
- paste and scissors

Students need to:

a. Color the shapes, being sure to press firmly with the crayons.
b. Cut out the shapes and paste them to the sheet of construction paper.
c. Add the face and hands to the child and the handle to the umbrella.
d. Add background details with crayons (puddles, clouds, raindrops, etc.).
e. Sprinkle DRY blue tempera paint on the picture.
f. Spray the picture with the spray bottle to create the drippy, rainy effect.
g. Set the picture aside to dry.

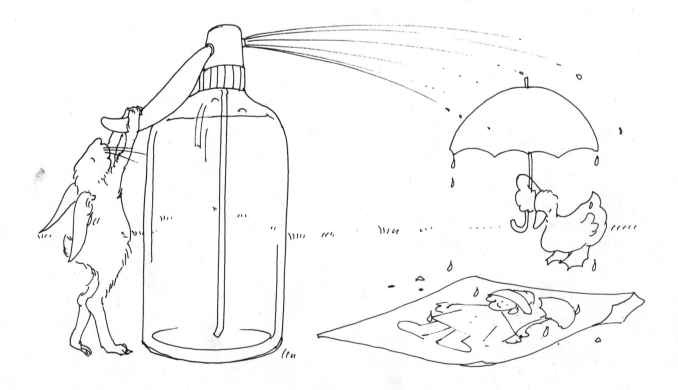

Slicker Kid Form

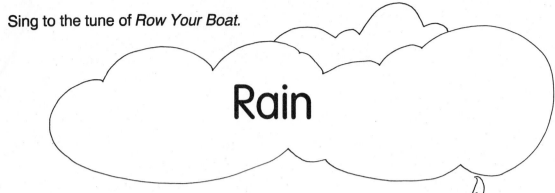

Rain

Rain, rain, rain comes down
From a cloudy sky.
Cools the herd and grassy plain.
The land no more is dry.

Cindy Lunsford

       Learning about Weather

Sing to the tune of *Frére Jacques* - sounds best as a "round."

Stormy Weather

Stormy weather, stormy weather

Dark black clouds

Thunder's loud

Then the rain starts falling

Raindrops softly calling

Splash, splash, splash

Splash, splash, splash.

Cindy Lunsford

Rain Cycle Wheel

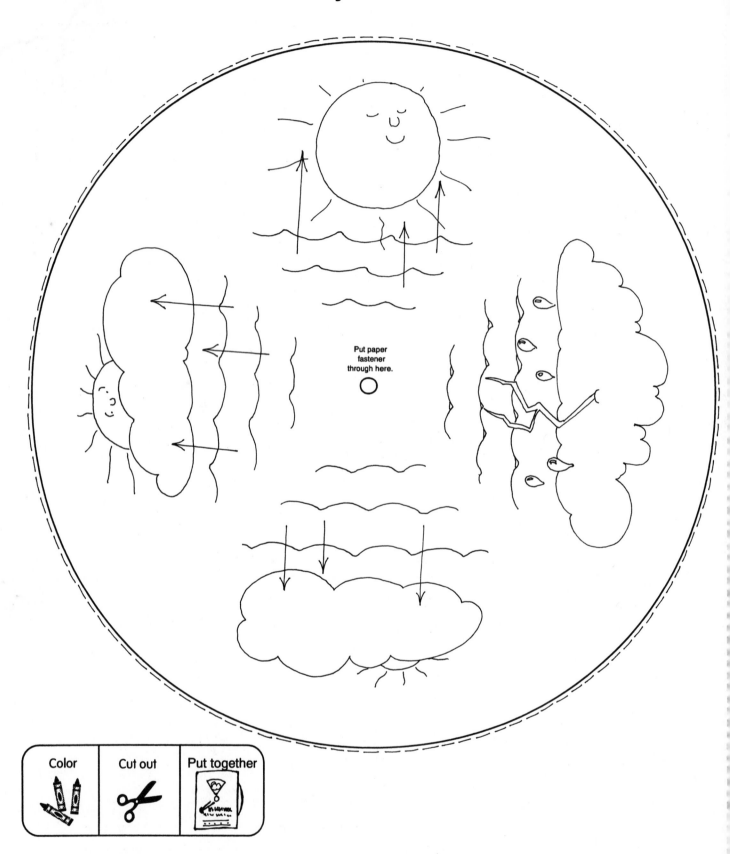

Put paper
fastener
through here.

| Color | Cut out | Put together |

Rain Cycle Wheel Holder

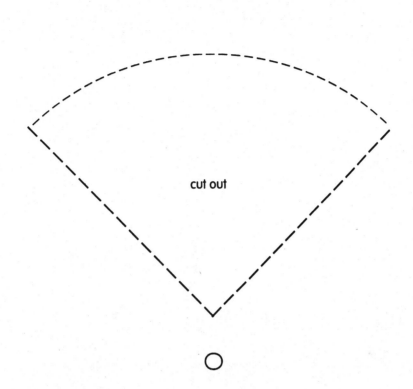

cut out

My Rain Wheel

Look and see the water cycle here.

Name _____

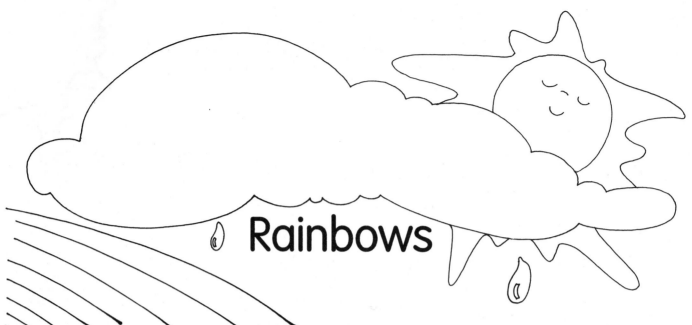

Rainbows

When rain falls down
And the sun shines behind,
You can see a rainbow.
It's easy to find.

Sunlight breaks apart
On the raindrops passing by.
Bright colors bend
And an arc fills the sky.

Jo Ellen Moore

Learning about Weather

A Rainbow Chant

Seven colors make a rainbow
Shining in the sky.
Can you name them?
Let's all try.

Red
Orange
Yellow
Green
Blue
Indigo
Violet

Seven colors make a rainbow
Bending across the sky.
Can you see them
Shining up so high?

Jo Ellen Moore

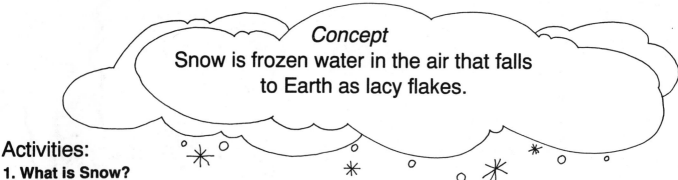

Concept

Snow is frozen water in the air that falls to Earth as lacy flakes.

Activities:

1. What is Snow?

a. Explain that snow is frozen water. Each frozen drop looks like a piece of lace. When the snowflakes land they cover up everything they touch and can form large piles of snow. Explain that sometimes the snow is blown back up into the sky over and over again. When this happens the snow begins to form balls. When these fall to the Earth we call them hailstones.

Read *Snow is Falling* by Franklyn Branley (Thomas Y. Crowell, 1963) to help children understand what happens to create snow.

b. Have children look for pictures of snow scenes in magazines, newspapers, etc. This is especially important for children living in areas where it never snows. Put these pictures into a scrapbook for everyone to share.

c. Scrape a chunk of ice to create very small pieces for children to feel, smell and taste. Explain that this is not snow, but it resembles snow. Ask them to tell you words that describe how the "snow" feels and smells. Ask them to taste the "snow" and tell you how it tastes.

2. Snow - A Poem

Teach your students the snow song on page 24.

3. Make Snowflakes

If your children are comfortable with scissors have them cut out simple snowflakes. Use lightweight paper squares and only fold them in fourths.

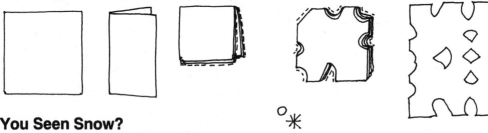

4. Have You Seen Snow?

Make a simple bar graph showing who in your class has or has not seen snow. Make a large form on a bulletin board. Give each child a small paper doily. Have children write their names on the doily with crayons. Have each child come up and place his/her snowflake in the correct column.

5. Magic Snowstorms

Each child will need:
- small glass jar with a screw-on lid (screw-top baby food jars work well)
- waterproof glue
- small plastic objects - miniature snowmen, people, houses, animals, trees, etc.
- small bits of white rock or sand
- glitter
- water (boiled, then cooled)

If you do not have the items mentioned above, use self-hardening clay to create mountains, trees and buildings. Add details with waterproof marking pens.

a. Turn the lid upside down. Put some waterproof glue in the center of the lid. Add a layer of white rocks or sand (mixed with glue) to form the ground for your scene. (Keep the rock and glue away from the edges of the lid or it will not screw back on the jar.)

b. Create a scene using small plastic objects (snowmen, people, houses, trees, animals, etc.). Glue the base of the pieces and set them firmly onto the rock base. Glue plastic objects onto the base. If you make things from clay, details can be added using waterproof marking pens. Let the glue dry before doing the next step.

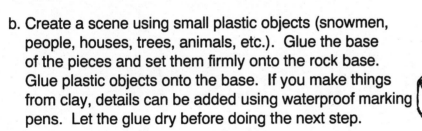
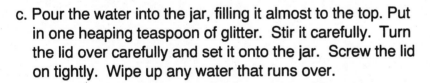

c. Pour the water into the jar, filling it almost to the top. Put in one heaping teaspoon of glitter. Stir it carefully. Turn the lid over carefully and set it onto the jar. Screw the lid on tightly. Wipe up any water that runs over.

d. Turn the jar upright and watch the snow fall!

6. Create Snow Pictures

Create "snow" pictures using whipped Ivory Snow Flakes detergent as snow. Put a box of Ivory Snow Flakes in a large container. Add a small amount of water. Beat the mixture with a hand beater until the soap has a consistency of thick whipped cream.

Give each child a sheet of blue or black construction paper and a dish of the soap mixture. Let them create snow creatures or snow scenes using the soap.

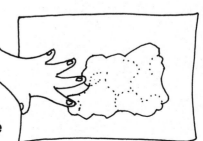

Have plenty of paper towels on hand for clean-up time.

Note: Check with parents before doing this activity to be sure no one is allergic to Ivory Snow Flakes.

Snow

Snow is falling all around,
all around, all around.
Snow is falling all around
Covering up the ground.

Icy flakes like bits of lace,
bits of lace, bits of lace,
Icy flakes like bits of lace
Are falling every place.

Jo Ellen Moore

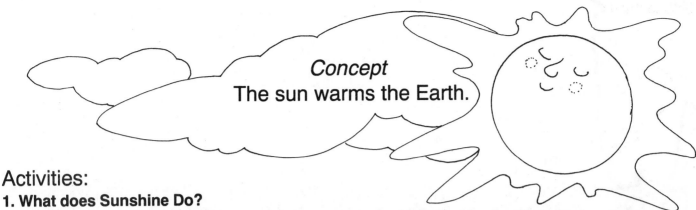

Concept
The sun warms the Earth.

Activities:

1. What does Sunshine Do?
Ask children to tell you what they know about what sunshine does for the Earth (makes it warm, helps plants grow, etc.). Explain that the warm sunshine has a part in making it rain. At this point review what they learned about the cycle of evaporation and condensation you talked about in the rain section (page 11). If you didn't make the rain wheel at that time make it now.

2. Sunshine - A Song
Teach your children the Sunshine song on page 26. Then make torn-paper suns.

Children will need:
- tag circles
- small scraps of construction paper in the "warm" colors (red, yellow, orange)
- glue

Have children select the colors they wish to use and glue them all over their tag circles. Encourage them to tear the pieces into interesting shapes. They may also want pieces of construction paper to stick out from the edge of the tag circle to look like sun rays. (Remind them to only put glue on the edge that touches the tag circle.)

3. "Fun in the Sun" - A Class Picture Book
Brainstorm to create a list of all the kinds of "fun" children can have on a sunny day. Have each child select one activity to paint. Discuss ways to make it look "sunny" in the picture. When the pictures are dry, take a black marking pen and have children either write or dictate a sentence about the picture. Punch holes and tie all of the paintings into a tag cover.

4. Sun Visors
Make sun visors. Reproduce the form on page 27 for each child on construction paper. Have children color and cut out the visor. Attach yarn to each side and tie in back to hold the visor on the child's head.

         Learning about Weather

Sunshine

Have you ever seen the sunshine,
the sunshine,
the sunshine?
Have you ever seen the sunshine
On a hot summer's day?

It gives light and warms you.
It helps trees and flowers too.

Have you ever seen the sunshine
On a hot summer's day?

Jo Ellen Moore

Learning about Weather

Sun Visor

Learning about Weather

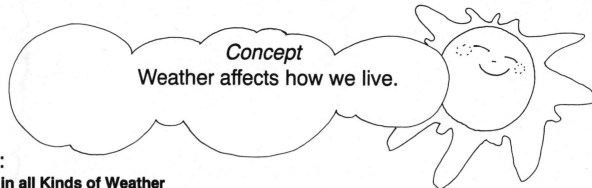

Concept
Weather affects how we live.

Activities:

1. Clothing in all Kinds of Weather

a. Help children understand that weather affects how we dress. Discuss how important it is to dress in a correct way for the weather outside. Ask children to tell how they would dress and why if they were going outside to play on a _____ day. Use the picture cards from pages 45 and 46 for illustrations.

    rainy, snowy, cool, breezy, hot, sunny

b. Read *Bear Gets Dressed* by Arnold Lobel (Harper & Row, 1986). Reproduce the bears on page 30 to use along with your class calendar. Each day check the weather outside and discuss what the bear should wear. Pin the appropriate bear picture to the day's date on your calendar.

c. Bring in a collection of clothing appropriate for different types of weather. (Look in yard sales and thrift shops and ask parents to contribute to the collection.) Try to include everything from a snowsuit to shorts and tee-shirts.

Hold up a piece of clothing. Have your students name the item and tell when it would be worn. With some items there will be more than one correct answer.

Put all of the clothing in a pile. Put cards with pictures and/or words naming a type of weather in a bag or box. Select a child to pull out one card. The child reads the card and selects the clothing that he/she would wear for that kind of weather. You might even have the child put on the clothing (over his/her own clothing).

d. Pin up four sheets of tag or butcher paper. Label them "rainy day," "snowy day," "cool day," and "hot day." Pass out catalogs and magazines. Have children find appropriate clothing for each type of weather. Paste the pictures to the correct poster.

e. "What Should I Wear?"
Have children select appropriate clothing for different types of weather using the forms on pages 31 and 32. Children cut out the clothing and paste it on the correct bear. Have them check their answers with a partner before pasting pieces in place.

2. Houses in all Kinds of Weather

a. Read a book about houses around the world to your students. Here are three you might try:

This is My House by Arthur Dorros; Scholastic, 1992
Houses and Homes by Ann Morris; Lothrop, Lee & Shepard, 1992
Houses of snow, skin and bones by Bonnie Shemie; Tundra Books, 1989 (This is part of a series on different types of native dwellings.)

b. Use any pictures of homes in various climates around the world to share with your students. (Or use the picture cards on pages 47 and 48.) Ask children to tell why we need to live in houses. Guide them to come up with answers about keeping dry, warm, etc. Ask if they know what kind of a house people might live in in a place that is very cold, hot, dry, wet, etc. Use the pictures of houses in other areas and climates to help move the discussion along.

A home on the plains.

A home in the tropics.

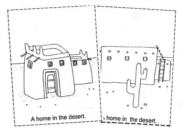

A home in the desert.   home in the desert.

A home in the mountains.

c. Reproduce the form on page 33. Have children mark the correct houses as you give these oral directions:

- Put a red X on the house you might see where there is a lot of snow.
- Put a black line under the house you might see in a hot desert.
- Put an orange circle around a house you might see in a forest.
- Color the house you might see in a place with a lot of water.

3. Outdoor Activities

a. Help children understand that weather affects what we can do outside. Discuss how the weather can keep you from going outside to play. (You may want to digress here and talk about what you can do indoors if you can't go outside to play.)

b. Brainstorm to create a list of all the kinds of things you can do during specific types of weather.

- after it has snowed - sled, build a snowman, make snow angels
- on a hot, sunny day - run through the sprinkler, swim
- on a windy day - fly a kite, play with a pinwheel
- on a rainy day - go for a walk with an umbrella, splash in puddles (with rain boots, of course!)

c. Reproduce the activity sheet on page 34.
Have children fill in the type of day, then
draw a picture of themselves at play.

Note: Reproduce these bears to use with the calendar activity on page 28.

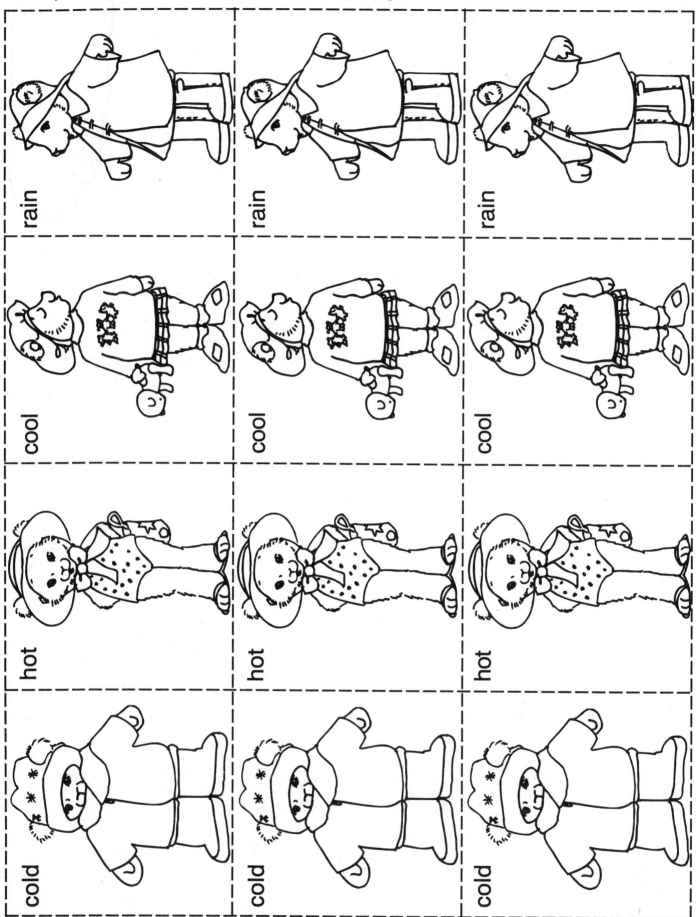

rain

rain

rain

cool

cool

cool

hot

hot

hot

cold

cold

cold

Learning about Weather

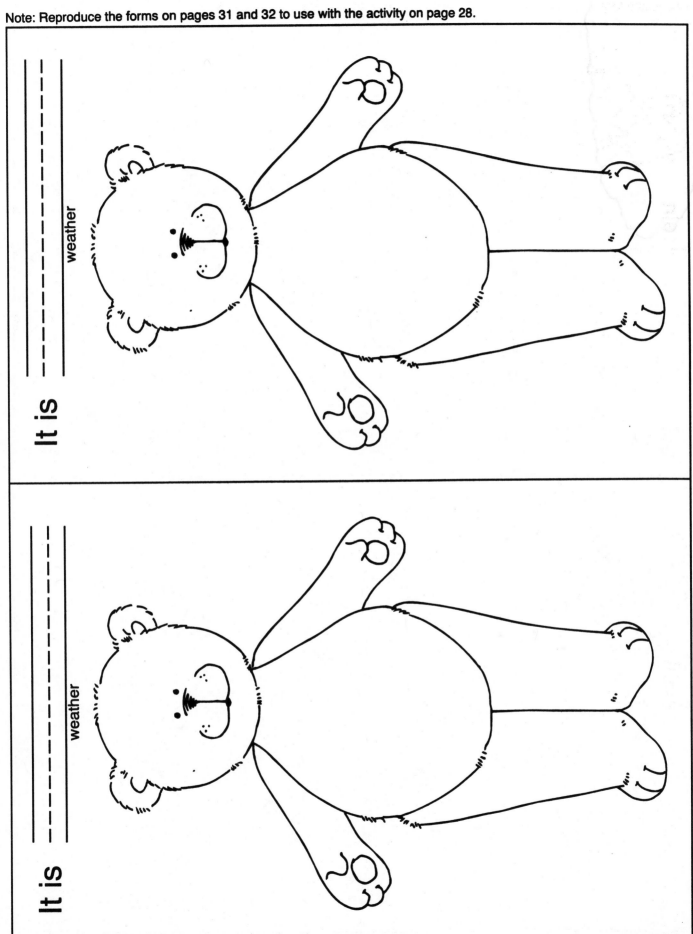

It is _____
weather

It is _____
weather

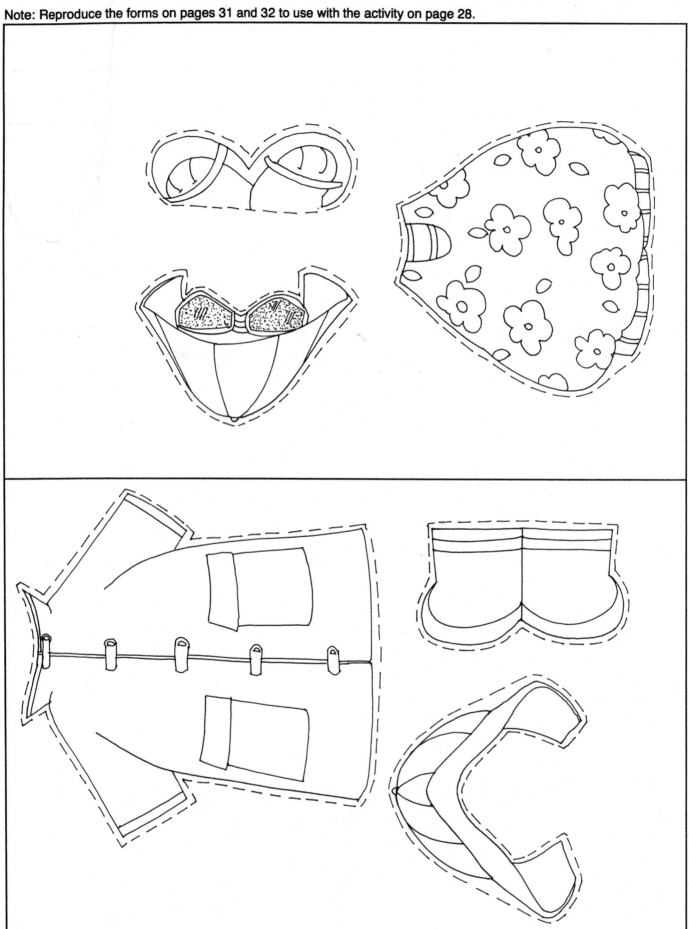

Name _____

Homes for All Kinds of Weather

Note: Use this form with activity 3c on page 29.

Name _____

What I can do on a _____ day.

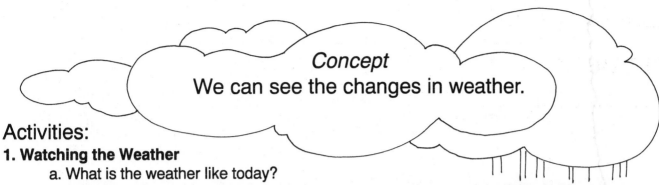

Concept
We can see the changes in weather.

Activities:

1. Watching the Weather

a. What is the weather like today?

Discuss what we see when we watch the weather (if it is raining, if the wind is blowing, if it is a foggy day, etc.). Have children look outside the classroom and describe the weather they see. Give each child a copy of the "window" form on page 37. Have them draw a picture of the kind of weather they saw through the classroom window. (Repeat this activity throughout the year selecting specific types of weather you want children to record.)

b. Set up a weather calendar in your classroom. Use a standard classroom calendar form (or make your own on butcher paper) with the reproducible weather symbols on page 38. Each day have your students decide which symbols best show what the weather is like. Pin the symbol on the calendar. You need to change the symbol during the day if the weather makes a big change.

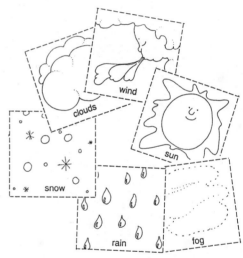

c. If you have limited space on your daily calendar, make a weather wheel to post next to the calendar. Use the pattern on page 39. Paste it to a circle cut from tag or lightweight cardboard. Cut the arrow out of tag or cardboard also. Attach it to the wheel with a large brass paper fastener.

2. Knowing about Weather

Why do people want to know what the weather will be? Discuss how knowing what the weather will be can help us. Guide children by your comments and questions to come up with answers such as:

• We need to know the weather so we will know how to dress.
• We need to know if it is going to rain so Mom and Dad will know if they have to pick us up after school.
• Farmers need to know so they can take care of their crops.

 Learning about Weather

3. Using Weather-Telling Instruments

a. How hot or cold is it? - Ask children what they know about thermometers. Someone in class will be familiar with having his/her temperature taken.

Explain that the weather has a temperature too. Show children a real outside thermometer (use the largest thermometer you can get). Explain that the liquid inside goes up when it gets hotter and down when it gets colder. Mark the top of the liquid with a piece of tape. Take the thermometer outside, wait a few minutes and mark the top point again. Ask children to tell you if the liquid went up or down. See if they know if that means warmer or colder.

b. Make a model of a thermometer using tag, wide red and white ribbon, a black marking pen, needle and thread and a safety pin.

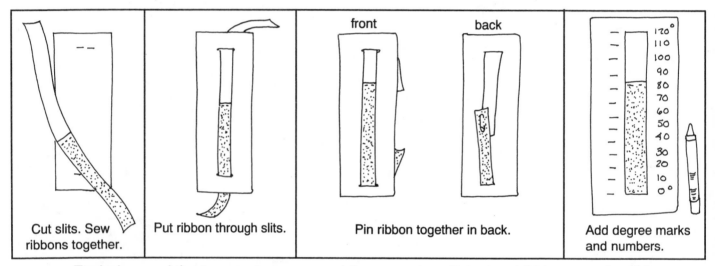

| Cut slits. Sew ribbons together. | Put ribbon through slits. | front   back Pin ribbon together in back. | Add degree marks and numbers. |

Each day read the temperature on a real thermometer. Select a child to move the ribbon up or down so the top of the red line marks the actual temperature. If your children cannot read numbers yet, move the ribbon yourself and have the children watch to see if it moves up or down. Then discuss whether this means hotter or colder.

c. Which way does the wind blow?

Make a weather vane. Take it outside to check which way the wind is blowing. Make a chalk arrow on the ground showing the direction. Check it several times during the day, drawing a chalk mark on the ground each time. At the end of the day have the children observe the chalk marks to see how the wind changes. (You may want to have each child make one to take home.)

For each weather vane you will need:
- a thread spool
- a drinking straw
- an arrow cut out of heavy paper
- a strip of tape

Put tape over the hole in one end of the spool.
Glue the arrow to one end of the straw.
Put the straw into the open end of the spool.
Set the spool outside and see which way the wind is blowing.

Note: Use this page with activity 1a on page 35.

Name _____

Look out the window.
What is the weather today?

Learning about Weather

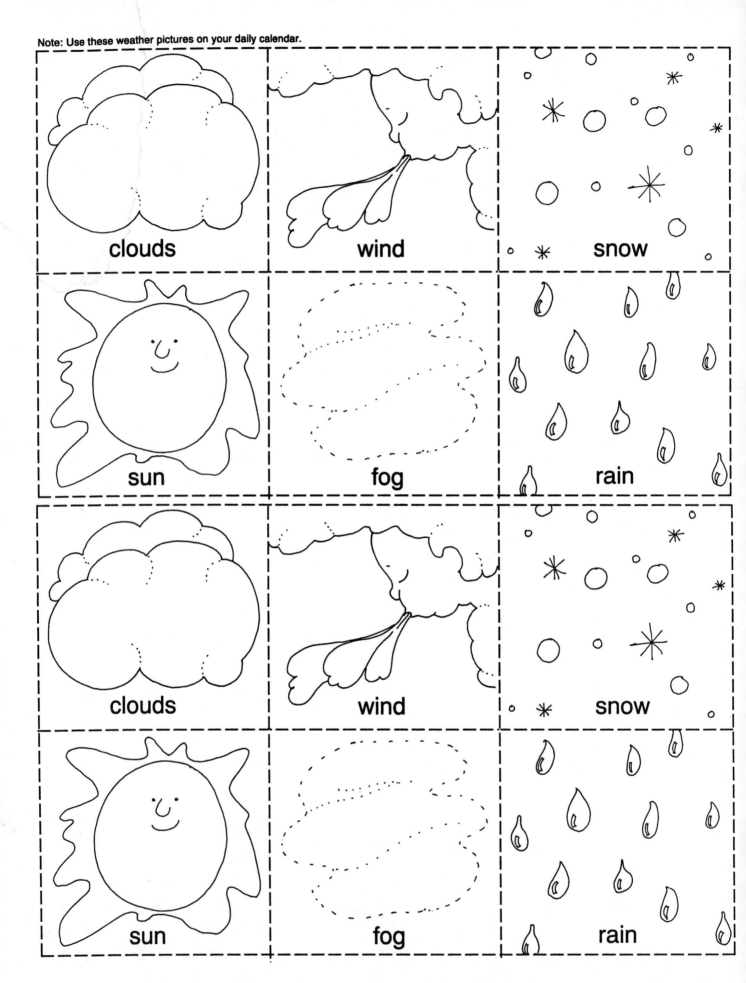

clouds

wind

snow

sun

fog

rain

clouds

wind

snow

sun

fog

rain

Note: Use the wheel with activity 1c on page 35.

Weather Wheel

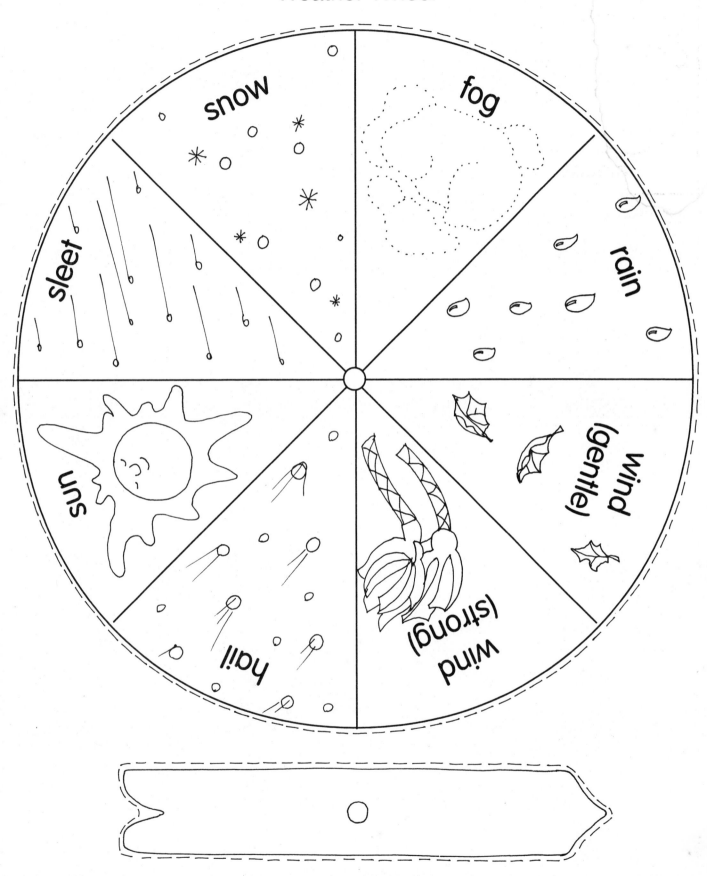

Cut two arrows for two choices.

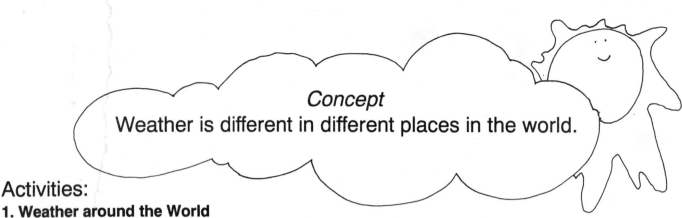

Concept

Weather is different in different places in the world.

Activities:

1. Weather around the World

Discuss how weather/climate is different in different parts of the world. You may need to ask questions such as "What is the weather like where we live? Does it ever change?" or " Have you ever lived in another part of the world? What was the weather like there?"

Pictures from books and magazines showing other parts of the Earth can help also. You may want to reread some of the books listed on page 29 to clarify this concept.

2. Cold Places - Hot Places

a. Show children a globe. Point out the polar regions. Explain that these areas are always very cold. Point out the areas along the equator. Explain that these areas are usually hot. Review how extremes of weather can affect how we dress and what we do. Ask children to recall how they might dress in the very cold parts of the Earth and in the very hot areas.

b. Reproduce the animals on the bottom of this page and page 41 for each child. Children are to decide which animal is dressed for which type of weather, then to paste the animals in place on the globe.

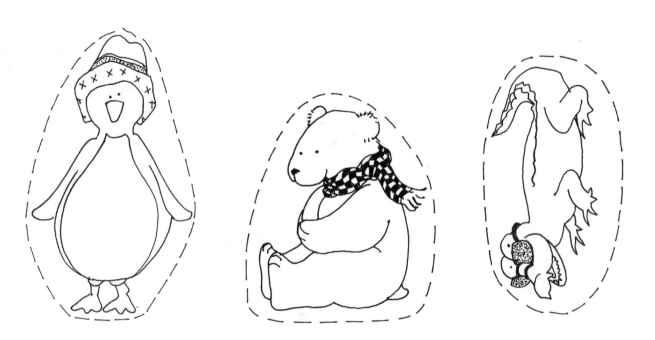

Name

Cold Places - Hot Places

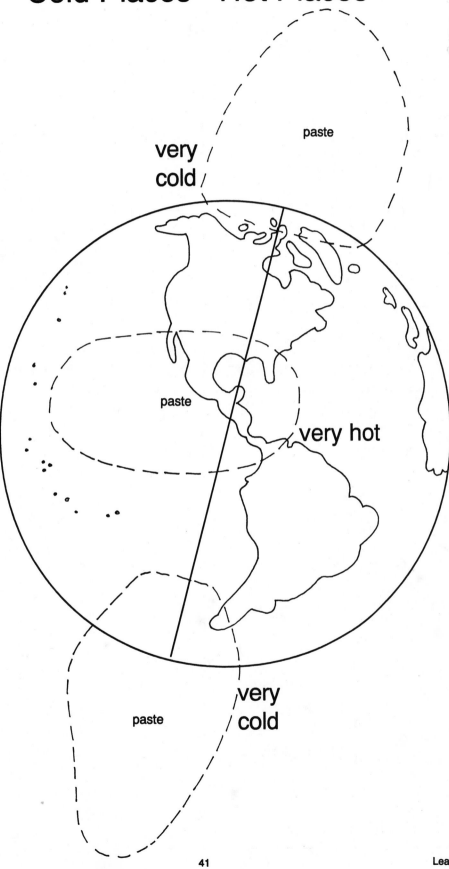

very cold

paste

paste

very hot

very cold

paste

Learning about Weather

Note: Use these picture cards for various activities in this book.

rain

snow

sunshine

fog

Note: Use these picture cards for various activities in this book.

sleet

hail

breeze

hurricane

44

Note: Use these picture cards for various activities in this book.

 Learning about Weather

Note: Use these picture cards for various activities in this book.

Learning about Weather

Note: Use these picture cards for various activities in this book.

a home in the tropics

a home in the desert

a home in the mountains

a home on the plains

Note: Use these picture cards for various activities in this book.

a home in the tropics

a home in the desert

a home in the mountains

a home on the plains

Learning about Weather